CW00342619

Image Filing Systems

A Practical Evaluation Guide

J A T Pritchard

PUBLISHED BY NCC PUBLICATIONS

British Library Cataloguing in Publication Data

Pritchard, J. A. T.
Image filing systems : a practical evaluation guide
1. Image processing
I. Title
621.38'0414   TA1632

ISBN 0-85012-516-2

First published in 1985 by:

NCC Publications, The National Computing Centre Limited, Oxford Road, Manchester M1 7ED, England.

Typeset in 11pt Times Roman by BPCC Northern Printers Limited, 76-80 Northgate, Blackburn, Lancashire; and printed by Hobbs the Printers of Southampton.

ISBN 0-85012-516-2

Preface

NCC established an interest in office automation in 1978 and published *Introducing the Electronic Office* (see Bibliography, item 1) in 1979. The work from which that book derived set the scene for a follow-up programme – and in 1980 NCC commenced a programme of work supported by the Department of Industry (now the Department of Trade and Industry) through the Computers, Systems and Electronics Requirements Board (now the Electronics and Avionics Requirements Board).

The objective of NCC's Office Systems Division is to help increase the efficiency and use of office technology by:

— keeping potential users informed of the significant developments that are taking place;

— disseminating the experiences of the larger users to a wider audience;

— offering guidance on the planning and implementation of office systems which utilise the new technology.

During 1982, which was designated Information Technology Year (IT82), NCC introduced four Information Technology (IT) Circles:

— Office Technology Circle;

— Communications Circle;

— Data Processing Circle;

— Security Circle.

A fifth IT Circle for Microsystems was added in 1985.

The Office Technology Circle holds an annual conference and interactive workshops, and publishes state-of-the-art reports. It helps to further the mutual exchange of information, leading to the generation of new ideas. Its area of interest embraces text processing systems, person-to-person communication systems and information management and access systems. Further details may be obtained from the NCC IT Circles Administrator (see Appendix D, Useful Addresses).

This book is one of a series of evaluation guides concerned with office technology products. Other guides deal with electronic mail systems, facsimile, viewdata systems, office printers, office workstations, text filing and retrieval systems, computer-assisted retrieval of microform systems, graphics in business, and intelligent photocopiers.

Acknowledgements

Assistance was given to the author by various organisations and individuals who gave their time for discussing products, development work, applications and ideas.

Thanks are due to:

Advent Systems, Wokingham

Correlative Systems International, Brussels

IBM

ICL Systems Strategy Centre, Bracknell

Kodak, Hemel Hempstead

Philips Business Systems, Colchester

Sintrom Electronics, Reading

Wang UK, Isleworth

Contents

Appendix

1 Introduction

AIM OF BOOK

This book is one of a series of evaluation guides concerned with office technology products. Each guide addresses a particular type of product. If an organisation's office automation strategy has a requirement for a specific type of product, then the corresponding NCC evaluation guide offers advice which will help in:

— preparing the specification for tender;

— evaluating the offerings from different suppliers;

— making a selection of a particular product.

Other NCC books and products give advice on the preparation of an office automation strategy (so that the right systems are identified), and on implementation and operational issues (to help to ensure that the anticipated benefits are achieved). Additionally NCC carries out office technology advisory assignments, in particular through the Snapshot consultancy scheme and the Technology Update scheme introduced in 1984.

WHAT IS 'IMAGE' INFORMATION?

In office technology, 'graphics' and 'image' may mean different things to different people. Confusion can and does arise. Graphics and image information are not the same, although there may be some overlap in the hardware and software components of the two types of systems. Sometimes, unfortunately, the terms are used interchangeably, or one is used when the other should be. The usage described in this book is in line with how these terms are most frequently used.

The five forms of information are data, text, voice, graphics and image. Together, they comprise the information resources of a nation, an organisation, a department, or an individual. Office technology investment is intended to lead to the more effective management of these resources.

Data, text and voice are familiar forms of information. Some examples of where they occur are:

— the input to data processing programs (data);

— the output from word processors (text);

— the spoken word (voice).

Graphical information, 'graphics', is information derived from stored numerical data, alphanumeric text and graphic elements (eg a polygon, a hammer and sickle or other national emblem, special characters used in mathematical texts, musical scores or maps, etc).

This derived information is presented (displayed or printed) in graphical form as histograms, pie charts, mathematical graphs, scatter diagrams, captions, cartoons, diagrams, etc. Graphical information is widely found to-day in:

— business microcomputers with graphics facilities;

— viewdata systems;

— certain television programmes such as news, documentaries and elections;

— teletext.

Image information, 'image', is derived from a scanning process. The input 'document' is scanned and digitised (ie represented as a sequence of bits – which are subsequently compressed using an efficient compression algorithm to reduce the storage space used and to reduce transmission times/costs). Subsequently, an image of the input document is derived and presented. It should be noted that the original document need not necessarily be paper, or even two dimensional. In some specialised image systems, it may be:

— a microfilm image (videomicrographics);

— displayed graphical information;

— a three-dimensional arrangement.

Whilst these definitions seem appropriate today, it is possible that they may need revision in the future as fifth-generation computer systems (FGCS) and knowledge-based information processing systems (KIPS) begin to emerge in the next five to 15 years. These systems may have facilities for pattern recognition, image interpretation (of still and moving pictures), natural-language understanding and dialogue, foreign-language translation, and other advanced capabilities.

IMAGE FILING AND INFORMATION RESOURCE MANAGEMENT

The subject of this book is image filing and retrieval systems. But what is image filing, and what is its position vis-a-vis the information resource management spectrum?

The market for image filing systems is not yet mature, but a number of systems for general-purpose office use (as well as some advanced systems for specific applications such as cartography, CAD/CAM and fingerprints) are available. Technological developments are bringing developments in the functionality of image filing systems, which in turn will make their cost much easier to justify. Some large organisations see a need for image processing in their evolving office automation strategy, and a few have reached tentative conclusions about how it fits in – as do some smaller organisations with particular requirements.

In many organisations, at this time, there is a great deal of experience of word processors, microcomputers and personal computers for document preparation, spreadsheet work, text filing and retrieval, and electronic mail. In other words, such markets are mature (although these markets and the products produced for them will continue to develop). The markets for local area networks (LANs), the emerging value-added network services (VANS), and teletex will mature or become more mature during 1985, due to better products, the continuing development of BT's networks and services, and the attention of users moving on from document preparation and information retrieval to communication of text messages and text documents.

Thereafter users will turn their attention towards voice filing and image filing, as some have already done.

Today an image filing system for office applications should, as a minimum, be able to:

— scan an A4-size black and white document;

— represent the digitised image, allocating one bit per pixel (prior to any compression);

— have indexing and retrieval facilities;

— present the image on a high-resolution screen and print it on a high-resolution printer.

It ought also to be able to:

— transfer the digitised image over a high-speed local area network;

— interwork with many internal and external equipments and services, including access to the random access (magnetic disk) mass storage of the organisation's mainframe computer and the transfer of images between similar equipments at different sites over wide area networks;

— permit images to be edited and merged with word processing text documents.

Subsequently, perhaps within the next five years, image filing systems will have a much greater degree of functionality than this, as well as being more cost-effective and easier to justify. (NB Some image systems currently available do have higher functionality – or will have shortly, when planned enhancements are released.)

This guide is particularly concerned with 'basic' image systems applicable to the scanning, filing and retrieval of paper documents that arise, and will continue to arise for several years, in general office work. The reasons for this are:

— many organisations have a need for a basic image filing system, whereas the market for 'advanced' systems is much smaller;

— organisations that acquire a basic system will want to integrate it with word processing, electronic mail and other office applications, with their LAN and other communications infrastructure components, and with high-resolution graphics workstations; indeed, the image system will be an integrated extension of existing office systems. On the other hand, organisations that acquire an advanced system may be able to justify it and use it on a stand-alone basis.

It would be impractical for an organisation to scan and digitise all of its existing and new hard-copy documents. A choice has to be made; each organisation must make its own choice. Some alternatives, when an image filing system is available, are:

— to scan and digitise the most important hard-copy documents;

— to maintain an electronic index of those hard-copy documents which are of a lower degree of importance;

— to have no electronic reference (ie keep a manual hard-copy index, or no index) for the least important hard-copy documents.

It should be noted that:

— the types of office documents which it is envisaged a basic image system would handle include:

— correspondence;

— handwritten notes, such as hand-drawn diagrams attached to completed insurance claim forms;

— printed documents with handwritten annotations;

— pages from books;

— cuttings from magazines, journals and newspapers;

— application forms for jobs, and other forms, completed in handwriting and/or typed text;

— documents with handwritten approval/authorisation signatures;

— in the electronic office, more and more documents will be

created in electronic form and be transmitted electronically, intra-organisation and inter-organisation, by services such as teletex; eventually these text services will evolve so that they can handle mixed-mode documents (ie text with embedded image);

— an image filing system would be of limited use if users can view, transmit and print a document, but cannot do such editing operations as select a part of the image. Users need to be able to embed in a word processing document items selected from one or several image documents, and to form new image documents with elements selected from existing filed image documents. (An image system without editing capabilities cannot allow users to manipulate its images, and the information filed therein is only a little more useful as a resource than information printed on paper – for all information held on paper is 'dead' information, because it cannot be manipulated. An image system, of course, can bring to life information on paper; once digitised, this information becomes potentially very much more useful);

— future image systems with image recognition capabilities will be able to recognise patterns (sequences and arrangements) of alphanumeric characters and graphical elements and may be self-indexing;

— although there will be a migration of documents from hard-copy generation to electronic generation, there are circumstances in which hard-copy documents will be preferred. These have been identified in *Electronic Indexing and Hard-Copy Management* (Bibliography, item 2), and include:

    — personal notebooks;

    — printout of electronic personal diary;

    — annotation of documents;

    — reading documents when travelling.

RELATION TO OTHER TECHNOLOGIES

In considering its requirements for an image filing system, an organisation should in particular review its existing use of, and need to use in the future, the following types of system:

— graphics;

— facsimile;

— photocopying;

— microform.

Graphics systems have an effect upon the high-resolution workstations and printers, and the transmission rates of local and wide area networks, that an organisation uses. An image filing system may use the same equipment and networks.

Fast facsimile transmission of document images may be an organisation's major or only need for the application of image technology. An organisation may not need an image filing system; facsimile may suffice. On the other hand, an image filing and transfer system may overcome shortcomings in offices that heretofore have used facsimile, but have not been able to file and retrieve the digitised images and display them at office workstations. Developments in facsimile machines and intelligent photocopiers may go some way towards meeting an organisation's image requirements. The justification of an image filing system requires that it shall not be used just as an expensive facsimile system and/or document photocopying system.

Microfilm and microfiche systems with computer assisted retrieval (CAR) may provide a satisfactory way of storing, indexing, retrieving and displaying images. In the USA, videomicrographic systems have been developed. In these the microfilm image is scanned and digitised; this forms the basis for the CAR microfilm system to be extended into an image filing system.

In conclusion, developments in facsimile, photocopying and CAR of microform may go some way towards meeting an organisation's image management requirements, and make an image filing system harder to justify. On the other hand, the development of image filing systems integrated with other office automation

applications, equipment, systems and services may pull an organisation towards investment in this technology.

APPLICATIONS AND BENEFITS

Image filing systems have been described as a technology looking
for an application, for a problem to solve. On the one hand, there
are many 'user' organisations, who are concerned about the
amount of paper flow and storage, the cost of paper retrieval and
handling, and the problems of making use of information that
exists on paper but is not digitised and stored for reuse in electronic
form, and who see image as a way forward. On the other hand,
suppliers have developed various types of image filing system that
are impressive, from a technological point of view, and compe-
tently handle the 'set piece' demonstrations in exhibitions and
show-rooms.

It is probably fair to say that the image market is immature.
Indeed, if image systems were given away (and they are not inex-
pensive), it is quite likely that many users would not know how to
make use of them: users are uncertain in what way systems and
procedures in office work might be reorganised to include image.
Indeed, image systems will become much more functional and
easier for users to justify. There will probably be a huge image
market over the next decade. At the beginning of 1985 at least one
UK user bought and installed an image system (for office
documentation); a small number of pilot systems are installed and
on trial, including the National Coal Board/Wang office automation pilot supported by the Department of Trade and Industry
(DTI); and several more systems are in the pipeline.

It is predicted that the following events will occur in the next two
or three years and bring the image market to maturity:

—   suppliers will get feedback from users and increase the
    functionality of image systems, as well as integrating them
    with other office systems, equipments and services;

—   value added network services (VANS) offering integrated
    image transfer and facsimile will appear, spurred on by the
    teletex service for the interchange of text documents
    (because users will want mixed-mode working), the con-

solidation of CCITT Group III facsimile standards (and the maturing of the Group III facsimile market) and the emergence of standards for Group IV facsimile and OSI. (NB It should be noted that in December 1984 DTI announced that work had been discontinued on Project Hermes as it was considered that the Hermes service would compete directly with, and discourage the growth of, database and other services offered by the private sector. Hermes was conceived as a teletex-based document-delivery service, which could have included image, teletex and mixed-mode documents);

— the number of high-resolution workstations and printers in use in user organisations for business graphics applications will grow and lead to users wanting inter-organisation image transfer facilities;

— several user applications of image will be identified.

Some of the current applications (and benefits) of image filing systems in general-purpose office work are:

— integrated mixed-mode document preparation ('electronic cut and paste');

— forms filling, and order entry and confirmation;

— project filing;

— signature and photograph recognition;

— correspondence management;

— compact filing on optical disk;

— archiving;

— insurance claims;

— 'fast action' applications;

— image enhancement.

Integrated mixed-mode document preparation refers to the situation where a letter, report, etc is being prepared by a word processor or personal computer. The text refers to 'illustrations', which may be existing hard-copy documents (or parts of them) such as:

— items in newspapers, magazines, brochures, catalogues, other reports and timetables;

— photographs;

— hand drawings;

— etc.

Each illustration, once scanned and digitised, can be electronically cut, or trimmed, and pasted into the text with a keyed caption – the text being formatted to accommodate the space for the inserted image. When the user is already using, or is able to justify, a high-resolution workstation for other applications, the addition of a scanner beside his desk, and the other apparatus to give the workstation image capability, can result in the following benefits:

— immediate preparation of mixed-mode documents without having to go to (and to wait to use) a photocopier, and without having to cut and paste the illustration manually into the printed text document;

— distribution of completed documents, for checking by authors and for end user recipients, electronically instead of in hard-copy form;

— availability of electronic documents for access and retrieval, editing and re-use.

In forms filling, a blank (ie not filled in) form can be scanned. Variable data can then be keyed into the image of the form when it is displayed. Forms which have been filled in manually can be scanned, so that an image database is built up and made available for retrieval, inspection and transmission of keyed and handwritten information. The forms could refer to job applications, orders, equipment maintenance reports, etc.

Project files normally include many handwritten notes and other documents which have not been produced by a word processor and/or which cannot be read by an optical character reader (OCR). With image filing, these documents can be viewed simultaneously by people at different sites prior to progress meetings.

Filed images of signatures and photographs can be retrieved for the purpose of authenticating a signature or the identity of a person.

Correspondence management is an application being studied at at least one UK pilot installation. Letters from members of the public and other organisations can be scanned, and their images can be cross-referenced.

The optical disk has an immense storage capacity – of around one gigabyte (GB) or one thousand million bytes per side. It has a higher bit recording density per square inch of recording medium than magnetic disk. It can save valuable office space, which is a considerable problem for some organisations, at the same time making electronic information available for inspection, transmission and re-use. In Europe, optical disk-based image filing systems are installed for the storage and retrieval of technical documents, correspondence, press clippings and land registration diagrams. In another instance, the electronic archive of digital data is being transferred from magnetic tape to optical disk in order to save office space and to avoid the cumbersome manual procedures for the maintenance of a tape library (eg rewinding).

In Washington DC, a pilot image filing project at the US Library of Congress employs optical disks. It is reported that a 100 disk 'jukebox' configuration gives a retrieval response time of up to 10 seconds for the display of document images upon the screen of 200 lines per inch high-resolution display terminals. Single-sided optical disks can hold between 10,000 and 20,000 images of documents of size up to 14″ by 11″. At this library, there are over 80 million items and a material acquisition rate of 10 items per minute. Anticipated benefits include:

— readers go direct to a free terminal, instead of having to determine which of four reading rooms to go to;

— document retrieval time, including form filling and document handling, is reduced from up to two hours;

— documents can be simultaneously read by two or more persons;

— reduced handling of and damage to hard-copy documents, especially high use and aged documents;

— savings in physical storage space in rooms;

— documents given indefinite archival life (images can be re-recorded on to new optical disks every few years, whereas print technologies cause documents to deteriorate and rot within 25 to 100 years.

Insurance claims involve handwritten notes and drawings, as well as policies, and typed text. This type of mixed documentation could be another application area, where the user wants to look up information.

Fast-action applications are ones where it is necessary to transmit urgent documents very quickly for action to be taken before deadlines are reached.

Image enhancement systems have advanced software for the analysis, alteration, editing and presentation of images of maps, charts, photographs and drawings which have many shades of grey in the original document. The term electronic 'painting' is sometimes used in this context. Application areas include pharmaceutical research, medical research, cartography, geology, metallurgy, graphic design, space research, aerial photograph analysis, architecture, typesetting, advertising, textiles design and quality control. The sceptical reader may say that facsimile, intelligent photocopiers, CAR (computer assisted retrieval) microfilm systems, OCR (optical character reader) systems, and other office technology can be used for several of these applications. Image filing systems can allow the user to carry out all of the functions of filing, retrieving, editing, transmitting, displaying and printing of digitised scanned documents, and can be integrated with word processing, electronic text filing and retrieval, electronic mail and other electronic office systems – all through the same workstation.

The image market has yet to mature. Identifiable application areas and tailored products, or general-purpose systems, may emerge. Systems may emphasise compact storage, integrated use with word processing and other electronic office systems, or image enhancement, etc.

The following diagrams are included to illustrate some of the features of image systems. (They are derived from printed output from various systems, which has been subsequently put through a photocopying machine and reduced.)

Figures 1.1 and 1.2 are derived from the same original document, but with different settings of the 'darken/lighten image' scanning mechanisms. Figure 1.3 illustrates the cutting and pasting of two graphs, which appeared in a newspaper, into a document containing keyed text. Figure 1.4 illustrates a drawing (of the layout of a bungalow). Figure 1.5 illustrates a part of a map such as might be used in a planning department or works department of a local authority or public utility, together with symbols which can be used for enhancing the image. Figure 1.6 illustrates a two-level road map. Figure 1.7 is self-explanatory.

Figure 1.1

Figure 1.2

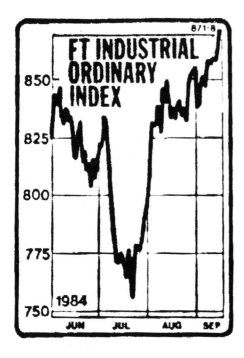

jk;gfj gfd j rgjm g
rjklfdbv m,.
cvxm,. cvxm,. xvcm, vcxzm, vcxm,.
vcxm,. vcxm,. cvxm,. vcxm,. xcvm,
. vcxm,. vcm,. vcx

fejlnfwe vfdlmnvcm, vcd ,.vfd vf
jkgfv sl nmvfd jkfdsj klgrs jol;gfdam
,. vfd m,. vfm
,. vdsm, sdm, sdcm,. dscm,. csdmcsdmdc m
df

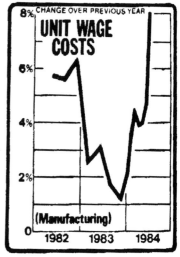

Figure 1.3

FLOOR AREA = 1,238 SQ. FEET
(Excluding Carport).

(Not suitable for Scotland).

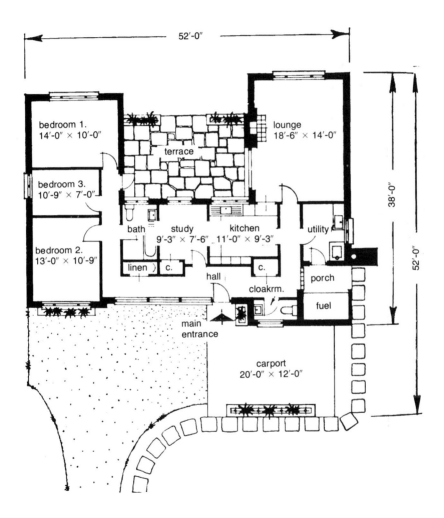

Figure 1.4

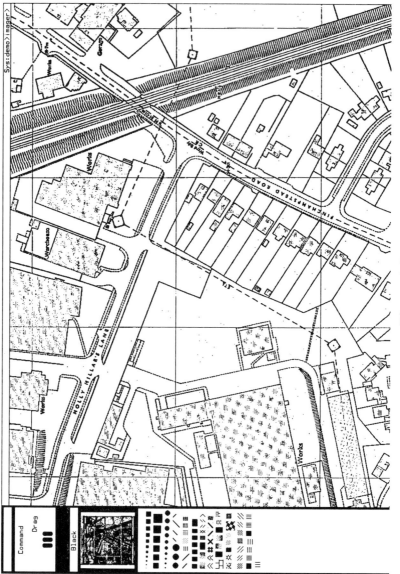

Figure 1.5

How to reach Advent by road:

Town	Distance
Reading	10 miles
Heathrow	20 miles
Central London	30 miles
Bracknell	5 miles
Maidenhead	15 miles
Basingstoke	12 miles
M4)J 10)	5 miles
M3(J 4)	10 miles

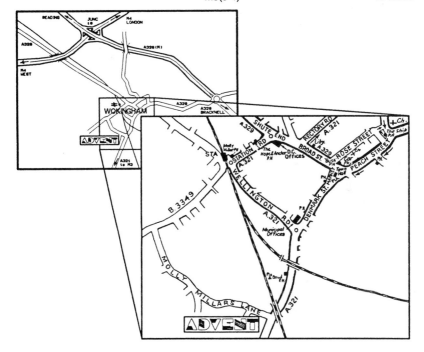

Produced on ADVENT DOCUMENTOR 20.
The map was produced via the ADVENT IMAGER 20.

Figure 1.6

This page was created from an original scanned image on the
DATACOPY MODEL 700.

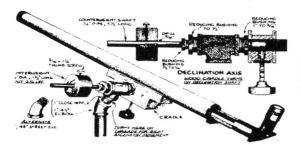

On the left is the full sized
image, scaled to fit into the
space as shown.

e picture to the right
a blow-up of a detail
the original drawing.
can be independently
aled on either axis.

e picture below is a
tated and inverted image
the detail drawing.

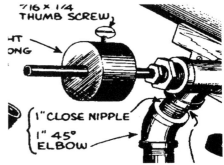

Note that the image can be placed
anywhere on the page to fit around
text.

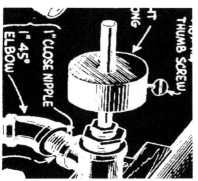

e drawing below shows the fine detail of image possible.

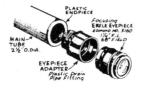

Image manipulated on an IBM PC
AT. Printed on a QMS Lasergrafix.

Figure 1.7

2 Criteria to Consider

INTRODUCTION

Evaluation criteria in this guide are considered under three headings:

— functional;

— ease of use;

— supplier.

Functional criteria are concerned with the facilities that may be offered by an image filing system. These criteria are not based on any specific system; therefore it is unlikely that any single product will include all the facilities listed. The prospective purchaser must assess the relative importance of each facility with respect to his specific requirements.

Ease-of-use criteria are concerned with the factors that determine the acceptability of an image filing system to its users within the office environment.

Supplier criteria deal with the supplier's capability to supply and maintain a product, and with the general terms and conditions of sale.

A checklist of all the aspects referred to in this chapter appears in Appendix A.

FUNCTIONAL CRITERIA

This section describes the functions which a user would expect to find in an image filing system, its integration with other electronic office systems, and security aspects.

Document Scanning

In the document scan phase of an image system, the input document is scanned by a camera to produce a stream of bits, which is compressed and written to disk.

The following aspects should be considered:

— type of document;

— size of document;

— nature of content;

— size of scanned area;

— resolution;

— number of bits per pixel;

— controls for setting up scanner;

— size of scanner;

— speed of scan.

The scanner may accept only original documents which are single sheets of paper. If a page of a book is to be scanned, a photocopy of the page would have to be obtained first. There may be a limit on the size of the input document (eg A4). If the document is too large (nb we are not, here, referring to the size of the area of the document that is of interest) and the document cannot be cut or folded, then either the document can be photo-copied and the photocopy cut to the right size or the document can be photoreduced. The scanned document may have colours or several shades of grey; whilst it can be scanned, the system may only be able to handle black and white, or black and white and half-tone (but a fewer number of shades of grey than the original) but no colours, so subsequent displays or printouts of the image may not appear like the original (and may indeed be useless).

The scanner may have controls which can adjust the size of the area which is scanned. For instance, a rectangular frame of light may be projected upon the flat bed upon which the document is placed. The frame can be adjusted so that it totally surrounds the document. At this point, it may be noticed that, in the 'vertical' direction, the document is shorter than the frame. Since the scanner scans one 'horizontal' line at a time and advances down the document in a 'vertical' direction from the top to the bottom of the document, considerable savings in storage space on disk can be achieved if there is a control to stop the scan when the bottom of the document is reached.

The scanning resolution or density may be 200 by 200. This means 200 points (or dots) per inch in a horizontal direction reading across the page from left to right, and 200 ppi (dpi) in a vertical direction (or, strictly speaking, lines per inch) reading down the page from top to bottom.

This gives about 4 million points for an A4 size page. This resolution should be adequate for office applications of image filing. Even quite small print is legible when viewed at this resolution, although it does not have a cosmetic quality. This resolution, however, is compatible with CCITT Group III facsimile standard. Higher resolution would increase the cost. A resolution of 400 × 400 would give a print quality indistinguishable from the original document and would be cosmetically attractive.

If one bit of storage is used for each scanned point, only two states (black or white) can be recorded. Some systems on the market use eight bits to represent each point. An eight bit analogue-to-digital converter digitises the scanned picture, translating each point into one of 256 possible digital codes representing different shades of grey. If a single bit is used, then half-tone original documents will be worsened. If three bits are used (as they are, for example, as a grey scale option in one product), then a half-tone original document will not be worsened as much as it will be if a single bit is used.

Controls vary from one system to another. There may be a control to adjust the size of the area to be scanned, and another to set the contrast level. Some examples are given:

— the scanner unit contains two fold-out lamps that provide the illumination necessary to capture the image. A scan can be initiated from either the keyboard of the workstation or the scanner. The scanner is adjustable to a height between 37cm and 58cm above the flat bed, and photographs vertically downwards. It is raised or lowered to adjust the length and width of the image. A light projects a frame on the original document lying on the flat bed to encompass the precise area to be scanned. When the scanner is at about the correct height, the final focus is made by rotating the lens to get a crisp image of the frame. A small toggle switch on the scanner head may be set for Text or Photo mode. In the former case a single bit of storage is used for each scanned point; in the latter case three bits of storage are used so that photographs and pictures can be stored in eight shades (levels) of grey. The system does not show the image on the screen prior to scanning, so that highlighting problems due to ambient lighting could be dealt with – specular reflection from the surface of the original document can sometimes 'white out' areas of the image, or create dark areas due to signal inversion (where the highlight intensity is great). If the scanner is located in a permanent position, where the ambient lighting has been appropriately adjusted and does not fluctuate, these problems can be avoided;

— in another system, the camera scans at up to 1,728 by 2,235 points or picture elements per page (high-resolution mode), with programmable resolution down to 1,728 by 1,117 points (low-resolution mode) per page. For the maximum size scan area of 8.5″ × 11″ a horizontal scan resolution of 203 dpi is provided, and alternative vertical scan resolutions of 203 dpi (high-resolution mode), 152 dpi (medium-resolution mode) and 101 dpi (low-resolution mode) are available. (NB In two of these scan resolution modes, the horizontal and vertical resolutions differ.) Scanning of any windowed area within the document can be controlled by the host computer (a personal computer). There are controls to allow the selection of the desired contrast level, from three options, for each maximum area (full page) scan or windowed area scan. There are two selectable half-tone patterns to choose from, matching the picture to a desired output quality. An integral illumination source provides the necessary light that is required. A lift top cover facilitates operation and is flexible enough to allow pages from books to be scanned. The scan time for a full page scan is 23 seconds;

— in a third system, there are four scan modes including high speed, continuous and slower incremental modes with programmable scan speeds. A high-resolution scan of 1,728 by 2,846 picture elements (4,917,888 picture elements) may be achieved. The scan resolution is programmable from this maximum down to 8 × 1 pixels. There are 8 bits of grey scale resolution per pixel, providing 256 shades of grey per pixel. The compact 3.3″ × 5.5″ × 5.8″ electronic digitising camera includes a focus and framing module that provides precise through the lens target alignment and visual focus aids without interfering with normal scan modes. This module projects a grid display through the lens onto the copy stand easel. It permits the camera to be repositioned, the object (source document) to be framed, and the camera to be focused in under 30 seconds typically. The base board of the camera stand is large enough to hold B3 size documents (and small three dimensional objects). Four quartz halogen lamps provide a flat illumination source. An optional hand-held controller is available for basic scanning functions to control the operation of the camera at the copy stand instead of using the host microcomputer. A 55mm flat fields lens, including an infra-red blocking filter, provides good quality images.

The term 'dynamic thresholding' is used to refer to the ability of a scanner to adjust automatically its black/white threshold, depending on the background. The level will vary across the page, as the facility exists for the background to be monitored on a local basis.

The scanner unit is not a piece of equipment which can be hidden away. It has to be easily accessible by the user and the camera requires a suitable lighting environment. Thus its size may be a consideration, as it may have to share desk top space with a workstation.

The dimensions of one scanner, which has the appearance of a camera on a vertical column which photographs vertically downwards to capture an object on a flat bed, are:

— height: 14″ (retracted), 23″ (extended);

— width: 12″ (collapsed), 24.5″ (extended);

— depth: 20.5″;

— weight: 20lb.

The dimensions of another scanner, which has the appearance of a photocopier, are:

— height: 110cm (43.3″);

— width: 70cm (27.6″);

— depth: 81cm (31.9″);

— weight: 85kg (187lb).

Those of a third scanner, which also has the appearance of a photocopier, are:

— height: 73.5cm (28.9″);

— width: 80cm (31.5″);

— depth: 50cm (19.7″);

— weight: 140kg (308lb).

In the scanning, or data capture, phase, the time taken determines how many documents can be added to the database in a day. This is not very important in some applications, but it is where the emphasis is on building a large database. In addition to document handling and controls setting up time, the system itself requires a few seconds to scan the document and transfer the image to storage.

In one system, the camera provides a visual field of up to 4.5 million pixels (1,720 × 2,592 × 8 bits) via a linear CCD (charge-coupled device) photosensor array of 1,720 sensor elements. This array is moved across the plane to capture up to 2,592 vertical bands. The picture is digitised by an 8-bit analogue/digital converter and is translated into one of 256 possible digital codes representing different shades of grey. A page is broken down into 200 ×

200 dots per linear inch in horizontal and vertical directions. Before the pixel is actually transmitted to disk, any discrepancies to the picture elements caused by interference are removed. The camera scans at 860 ms per line, and takes from 2.5 to 15 seconds for a complete scan, depending on the scanning speed. The transmission of output signals from the camera is broken down into three separate operations, the signals being in the form of time-multiplexed differential digital values. The first part carries one half, four bits, of the picture element brightness information; the second carries the four remaining bits; and the third carries information relating to the camera's status and control signals.

The transmission rate can vary from 2MB/sec to 58KB/sec depending on the scan time set by the user. The lower rate allows the scanner to interface with low speed storage disks.

The scanner in another system takes about one second to scan an A4 size page. It designates each scanned point as either black or white, although the user can adjust the scan contrast to one of 12 levels. A speed of 20 pages per minute can be achieved through a semi-automatic paper feed, which allows pages to be adjusted automatically or inserted manually.

In another system, the camera's scan speed is adjustable via a potentiometer that is accessible from outside the camera cover. The user can adjust scan times from 2.6 seconds to 19 seconds per 2,592 line frame. A high-speed interface delivers image data at rates up to 2MB/second.

In yet another system, the camera is configured for different applications via the selection of data rates by programming of four scan modes and setting scan area windows within the maximum scan area. For finer control, scan speeds are programmable within each scan mode. In continuous scan mode, the array moves continuously across its defined frame. There are also two incremental scan modes controlled by the camera clock and one by the host microcomputer. Scan times range from nine seconds per frame (continuous mode) upwards.

Some systems may offer a range of scanners with differing performances for different applications, and the ability to reduce or enlarge documents being scanned.

Image Storage and Filing

The second part of the data capture procedure, following the scanning of the original document, is the storage of the digitised image, together with associated indexing information (ie 'filing' the image), so that it can be accessed and retrieved subsequently.

The following aspects should be considered:

— storage medium and its capacity;

— location of storage medium;

— compression algorithm and speed of compression;

— indexing the image.

Any storage medium can be used in theory. However, image takes up much more storage than character-coded text. The application will determine how many images are likely to be stored at any time, how frequently filed images are accessed, and how fast a response time is required. These parameters, in turn, influence the choice of a cost-effective storage medium – or media, because (depending again on the application) different media may be appropriate for different categories of image.

Typically an A4 size page, scanned at a resolution of 200 ppi by 200 ppi, gives rise to about 4 million pixels and so to 4 million bits – or more, if multivalued pixels are used for shades of grey or colour. Typically a 10:1 compression ratio may be achieved. A page of character coded text requires, say, 3,000 bytes of storage. This gives a ratio before compression of about 170:1 for the storage of an image compared with the storage of a page of text generated by a word processor, or a ratio of 17:1 after compression. Another supplier estimates that an average image page requires over 60K bytes after compression, and sometimes as much as 120K bytes. Thus a floppy disk will not hold many images (perhaps seven), while a 10MB Winchester disk drive holds about 100 images. For some applications, floppy disk will be a perfectly adequate storage medium, just as it is in many word processing applications. For the storage of very large image databases, it may be advantageous to use the emerging technology of optical disks. Some systems have, or plan to have, on-line access to the mass storage capacity of an organisation's mainframe computer. (Further information about optical disk storage can be found in *Introducing Electronic Archiving*; see Bibliography, item 7.) One system uses video cassette recorders – a single VCR can store 90,000 images – or magnetic disk as the storage medium. Another system stores microfilm images in a central file. (When a request is received to view an image, an automatic loader/scanner, which uses computer indexing and robotics to control the centralised film file magazine selection and image retrieval, selects the film magazine and image required. The film image is scanned and digitised. The bit stream is transmitted over a local area network to the high-resolution image workstation, where it is stored in a local buffer and displayed. The same workstation can also display character coded text on its screen.)

The storage medium may be an integral part of the image system, eg floppy disk or Winchester disk. It may be necessary to store some of the images in another room or building, and it may be necessary to transmit images from one system to another. In these cases, the software must allow the image transfers to take place, and the local area network or other transmission channel must be able to cope.

The user does not need to know the mathematical details of the compression algorithm, but should find out what compression ratios are achieved for different types of document, such as typed text, printed text, handwriting, photographs, etc. Indeed some suppliers are reticent about disclosing their compression/ decompression algorithms and how these are implemented, but one uses Huffman compression.

In image systems, the trade-offs which have to be balanced by the supplier and by the user are:

— compression ratio;

— speed of decompression;

— quality of image.

A high compression ratio is good because it reduces the amount of storage used, and bad because it increases the time taken for decompression. A fast speed of decompression is required to give a fast response time when an image is being displayed or printed, but will increase the cost. The achievement of a good quality image, when displayed or printed, will add to hardware and software costs. Higher resolution will increase costs and memory requirements, but may also increase the time for decompression and the response time. Rather like motor cars, you want an image system that matches your user requirements, rather than one that is better but has an unnecessarily high degree of functionality and quality, or one that is better value but does not have sufficient functionality.

Suppliers' claims for average compression ratios are often in the range between 10:1 and 15:1, based on a spread from say 2:1 up to 40:1. In one-dimensional compression, individual scan lines are compressed in isolation. Two-dimensional compression goes on to look for similarities between neighbouring scan lines and may put the compression ratio up to 20:1. Compression speed need be no faster than the operation of the scanner (whereas decompression speed should be faster to give a low response time).

Generally, compression techniques for maps, engineering drawings, X-ray pictures, etc, in which subtle shades of grey are important and where the pixel is multivalued, are different from, and do not produce such a good compression ratio as, those for a typical office document, which is A4 size, black and white with no colour, with typed, printed and handwritten text and line diagrams.

The filed image must be indexed so that it can be retrieved. It is most unsatisfactory for the user simply to give each image a short unique name, even if there are only a few dozen images filed in the image database. Human nature being what it is, this can all too easily happen, as experience with text filing and retrieval shows, unless there is software to aid the user to perform indexing! Indexing software can be menu driven with a form displayed for the user to key in indexing information. For ease of retrieval, depending on the application, there may be several fields; many of these may be optional so that the user is not forced to spend too much time upon indexing when the document and the application does not merit it. Compulsory 'over-indexing' is burdensome and counterproductive for it tends to put users off using an electronic office system.

Indexing newly scanned documents must be done, but attention must also be given of course to index maintenance. Edited images must be indexed, cross referencing should be facilitated, the index should be able to be updated if the filing system is restructured and when images are deleted, and so on.

If indexing aids are not provided, users must nevertheless maintain an electronic index, although this is time consuming. People can soon forget what a short name refers to and, since a major advantage of office automation is shared access filing, difficulty (at best) and non-use (at worst) result if a group of users follow their own indexing conventions. Common indexing conventions should be used by members of a group, and this is facilitated if automated indexing aids are provided. It is suggested that as a minimum a filed image should be indexed with a short name, a short description, date of filing and name of person who filed it.

In one system, indexing information is held on floppy disk whilst images are stored on optical disk. In another system, indexing information is stored in a computer database and images (captured through a document microfilmer or a microimage processor) are stored in a 16mm cassette microfilm library. (In this instance an autoloader uses computer indexing and robotics technology to automate magazine selection and image retrieval. The retrieved microfilm image is then scanned and digitised, and could be reindexed if the digitised information is stored on magnetic/optical disk.)

Image Retrieval and Display

The next function is the retrieval and display of an image. Image retrieval is similar to text retrieval. The main difference between text display and image display is that an image can be displayed at different levels of detail (refer to Glossary for the meaning of 'Resolution'), whereas alphanumeric characters are displayed in a fixed way (for a particular mode of operation for a particular display equipment), eg 80 (or 40) characters per line, 24 lines per screen.

The following aspects should be considered:

— retrieval of image;

— display modes;

— display options.

The indexing of the digitised images should have been carried out with retrieval in mind. In some systems the user can, from the same workstation, interrogate an image database and also a database of text documents. In this situation, it may be desirable for the two databases to have integrated indexing/retrieval procedures so that a single enquiry will search the integrated database. The user may expect to be able to search the image database or integrated database/document library by such attributes as document name, document class, date of creation, author, subject, keywords (search terms), access codes, and so on.

In some systems, images may be retrieved by content, as opposed to retrieval by descriptive attributes. The logical power of the host computer can be utilised to search or match images, or to make comparisons between images.

Normally, it is not possible to display the entire document at the same level of detail at which it was scanned. In one system, the scanner and the display resolution are both 200 ppi × 200 ppi, but often the display resolution is only 100 ppi × 100 ppi. One supplier, whose high-resolution display shows 2,280 lines (each of 1,728 pixels), is planning to bring out in 1985 a lower-resolution display at one-third of the price. It will display images at a resolution of 100 ppi × 100 ppi and will be able to display text documents too. Later, in 1986, it is planned to bring out a display unit at 200 ppi × 200 ppi resolution that can display separately both image and text documents.

To illustrate the display capabilities that can be provided, and which a user might expect, one system offers the following display modes:

— 'view whole document' display mode;

— 'half-size scale' display mode;

— 'full-size scale' display mode.

In *view whole document* mode, the smallest scale mode, the height of the document is compressed (not to be confused with 'compression ratio', referred to elsewhere) and an entire A4-size original document can be viewed. The detail is sufficient for the user to identify the document, and if it is one which he wants to look at in detail he can pass on to another display mode. In *half-size scale* mode, the whole width of an A4 document and between one-third and one-half of its height is displayed. Most office documents can be read at this degree of resolution. In *full-size scale* mode, about one-twelfth of the original document can be displayed upon the full screen, but it is displayed in full detail, ie at the level of detail at which the original document was scanned.

In another instance, 'view whole document' display mode is referred to as 'full-view' mode (do not confuse with 'full-size scale' mode, which is referred to as 'full-resolution' mode).

Display options available on various systems include:

— pan (scroll);

— window;

— rotate;

— video inversion (reverse);

— binary/grey scale.

The ability to pan (scroll), vertically or horizontally, across an image (displayed in full-resolution mode) is essential. When the panning option is employed, a portion of the image is displayed on the full screen. Windowing or zooming, on the other hand, allows the user to identify an area of the image when it is displayed at less than full detail, to enlarge it to full detail and re-display it superimposed upon the existing display. This feature, also called spyglass, is – like panning – indispensible in most image filing applications, where a high-resolution screen which can display an entire image in full detail is not available. (Zooming is a most impressive feature when one sees it for the first time.) A variant of zooming is the enlargement of selected lines of text to make them legible; this feature is called 'magnifier bar'. Image processing has its own jargon and a term means whatever a particular supplier wants it to mean!

The option to rotate an image through 90° at a time is useful if the original diagram was not put into the scanner the right way up; indeed some documents do not have a top and a bottom, and the user would like to be able to see what different orientations look like. Video inversion is provided on most systems.

In multivalued pixel systems, there are facilities for the user to select the scan mode and the number of storage bits per scanned point. Likewise, there are facilities for the user to select the display mode, eg 'binary' for black/white original documents, and 'grey scale' display mode for originals with shades of grey.

Image Editing

In some applications, the editing of retrieved images is less important than the ability to look at a display or printout of a diagram, handwritten notes, photograph, etc, or some other feature (function) of the system. In many applications in the office or in specialised applications, the most important aspect of all may be the editing functions of the image filing system.

The following aspects are considered, the last of which – image enhancement – deals briefly with a number of specialised features which may be outside the requirements of a user organisation concerned solely with office documents:

— image manipulation;

— electronic cut and paste;

— image enhancement.

Image manipulation is the ability to derive a new image which is filed for retrieval and re-use, from an existing image (or more than one). An area of an image may be windowed, as in selecting an alternative display option, but for the purpose of transferring the contents of the window only into a new electronic document. Not only may areas of one image be defined for inclusion in a new document by windowing or other editing controls, but also areas may be defined for deletion. So an image may be edited, for example, by a number of deletions and substitutions, or a new image built up by copying into it marked/windowed portions of others. Other manipulations include moving, and rearranging the layout of, defined areas within an image.

But image manipulation, as described above, is of limited use on its own in office applications. A most important function in many applications is the ability to create mixed-mode documents from stored text, keyed text and stored manipulated images. The term 'cut and paste' or 'electronic cut and paste' is frequently used to describe this function. Some systems currently available have this feature, while others do not. The merging of word processing with image filing adds value to image manipulation and may have implications for printers, and local and wide area networking if a word processing system is being enhanced to a mixed-mode image filing/word processing system.

Some image enhancement features will be mentioned. The area of the image to be enhanced (or analysed or processed) may be delineated by a square, rectangular, circular or user-defined window.

Black and white areas within the windowed area may be calculated, based upon a user-defined black/white threshold level. Lengths of boundaries (perimeters) of objects within the windowed area may be calculated. Any point may be defined as the origin of an XY co-ordinate system, so that the co-ordinates, and distance from the origin, of other points may be determined. The definition of the window and of individual points is controlled by cursor movements, which may be keyboard controlled, with single stepped refinements for points and small object areas.

There may be facilities to:

— create a variety of overlays to be superimposed upon a given image;

— create black/white display images from multi-memory plane images using half-toning and edge enhancement;

— apply rectification corrections to geometric distortions such as paper stretch in the original document;

— do electronic painting (creating image, from keyboard and mouse control, onto a text document);

— merge text, graphic and image together by 'OR' or 'XOR' logic operations;

— do raster image to vector format conversion and vice-versa;

— convert alphanumeric text images into ASCII data format for subsequent text editing (character image recognition);

— provide an indication of average light intensity over the scanned object (light meter);

— provide an indication of grey scale distribution across a given scan line (image line histogram).

Image Transmission

The next decade will see a growth in electronic publishing and electronic document distribution, two important fairly new application areas of new technology. Transmission facilities in image filing systems will grow too. The growth in any of these three areas and in facsimile too will spur on growth in the others. Developments in British Telecom networks and services, and those of its competitor Mercury Communications, and in cable television and satellite communications, progress in CCITT and ISO standards work, and the emergence of new value added network services for image and mixed-mode transmission will all play their part in making organisations and society more information management aware, and less paper based.

At the present time (1985), the growth and integration of these systems, services and networks are just round the corner. Suppliers of image filing systems are aware of ways in which the functionality of image transmission in their systems could be developed, but details and dates are not yet available.

Prospective users of image filing systems should, therefore, watch out for the evolution of their corporate office automation and communications strategies, and suppliers' marketing strategies towards a compatible goal. Two questions that might be addressed are:

— do we want an image filing system that can interwork with facsimile?

— do we want an image filing system with mixed-mode document preparation facilities that can interwork with a mixed-mode service that evolves from teletex?

The transmission functions of currently available image filing systems may be considered from the aspects of:

— local options;

— remote options.

The provision of transmission functions, where available as an option, may increase the cost of a basic image system as an enhanced control unit, and additional software may be required.

Locally, a 10M bit/s Ethernet local area network or other local network may be used to interconnect several image workstations, or to allow the workstations to interwork with an office system and/or mainframe computer.

In one local configuration, multiple workstations can share access to filed images, and can share the use of scanners and printers. File access controls are provided, and the network speed is 2.5M bit/s. In another configuration, workstations can be connected to the supplier's electronic office system. An image workstation can interwork with the system, and with other image and non-image workstations. An image workstation can emulate a non-image workstation, as well as sharing access to image files with other image workstations.

Some systems can interwork with similar systems at different sites either directly or via a host computer. An image workstation may access a large image database held centrally at a remote computer, or a mixed database of text documents and image documents (or mixed-mode documents). Alternatively it may access a remote image database, or emulate a personal computer and access a text database, send/receive electronic mail, etc. In one particular system, the communications protocol for the transfer of images between different sites requires synchronous modems, switched or leased lines, and a minimum transmission speed of 4,800 bit/s.

Document Printing

In the document printing phase, the main functional consideration may be the resolution. Note that this does not necessarily have to be exactly the same as the scan resolution or the display resolution.

The following aspects should be considered:

— resolution;

— speed;

— mode(s);

— paper handling and other physical characteristics.

A print resolution of 200 ppi × 200 ppi may be adequate for many office applications. In some image systems, the user has a choice of several printers and the print resolution may be a factor in preferring one printer option rather than another. For example, one system offers the user a 200 ppi × 200 ppi thermal printer or a 300 ppi × 300 ppi laser printer.

Printing speeds are quoted in a variety of ways, for example:

— nominal speed is up to 12 pages per minute, with the exact speed depending on the complexity of the document (a laser printer);

— six A4 pages per minute (a thermal printer);

— prints and stacks a page in less than a minute;

— 20 pages per minute.

An image printer may be capable of working in more than one print mode, for example:

— printing digitised compressed images; the decompression may be performed in a special high-speed unit;

— as a photocopier;

— printing alphanumeric text characters produced, for example, by a word processor;

— true mixed-mode operation; for example, following the preparation of a mixed-mode document by electronic cutting and pasting, the printer in one pass prints the text in letter quality format and the embedded image at 200 ppi x 200 ppi resolution;

— printing images received from a facsimile machine/service.

There may also be a number of physical characteristics meriting attention.

One printer automatically detects the length of a printed document and cuts this off from a roll of electrostatic paper. Another uses specially treated thermal paper which is fed through the printer from a roll; the printed sheet is torn off against a bar. This thermal printer prints at 200 ppi × 200 ppi at a speed of about five pages a minute, depending on the density of the image. It costs about £1,500, whereas an optional laser printer costing about £22,000 is available. It prints at up to 12 pages a minute and software enhances the 200 ppi × 200 ppi resolution, at which the original document is scanned and stored, by calculating intermediate pixel densities so that the final document is printed at the higher resolution of 300 ppi × 300 ppi. The laser printer not only gives a higher quality printout, but supports several paper sizes, including A4, 8.5″ × 11″, and 8.5″ × 14″, supports a variety of character sets including Courier, Gothic, Geneva, Prestige Elite and Boston, and has dual paper trays that hold 250 sheets each, providing a maximum output capacity of 500 sheets.

Another supplier is bringing out a printer with additional paper handling facilities – a four bin feed and a 20 bin sorter.

Size, portability, noise, power requirements and other factors may be relevant considerations. The dimensions of a thermal printer weighing 12 lb (5.5 kg) are 4.4″ (height) × 11.4″ (width) × 16.5″ (depth), and of a 285 lb laser printer are 36″ (height) × 26″ (width) × 25.8″ (depth).

The reader's attention is drawn to another NCC publication in this series of practical evaluation guides – *Office Printers* (see Bibliography, item 5) – and to NCC report *Business Graphics*, pages 34-37 (see Bibliography, item 6).

In addition to printing the document on paper, facilities are available in some systems for output onto slides and films.

System Controller and Integration

The system controller, or host computer, may be a microcomputer or a minicomputer. It co-ordinates the operation of the system's components. In some applications, it is important that the image filing system should interwork with other internal and external systems, services and equipment.

The system controller may have software to support mouse-controlled editing operations, as well as software which integrates image filing with word processing (for electronic cut and paste), database, graphics and communication software.

Figure 2.1 illustrates a typical image system with the system controller interacting with the scanner, interface compression unit, mouse, high resolution display screens, disk storage, printers, Ethernet LAN, general-purpose interface bus (GPIB) and RS232/V24 wide area communications interface.

A useful integration feature is character image recognition. Segments of scanned documents containing alphanumeric characters can be selectively converted to ASCII code. This reduces storage requirements and allows the text within the image to be modified by word processing, or combined with other text. Software can automatically separate the text areas from the rest of the image, or the user can define windows to restrict the text areas to be analysed. The software is able to identify characters in one of two ways. It may be pre-programmed to recognise certain fonts. It

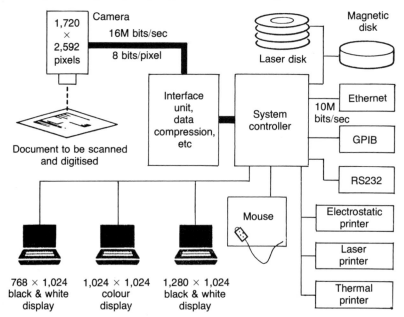

Figure 2.1   An Image Filing System Configuration

may also have an additional interactive learning capability, so that when a new font is to be learned a sample is scanned. The user controls the accuracy of this character recognition process. Not only can the text, which has been recognised and converted to ASCII, be integrated with word processing software, it can also be integrated with database management systems and other software.

Some well-known suppliers of computers and office systems also supply image filing systems that can be integrated with these systems.

The IBM 8815 Scanmaster1, for example, can use the DISOSS document distribution service (DDS) and document library service (DLS), and can complement tasks performed by DOSF, Displaywriter, 5520 Administrative System and the 8100. The Image View Facility (IVF) operates in conjunction with DISOSS/PS (Distributed Office Support System/Professional Support) and GDDM (Graphical Data Display Manager). It uses the mail log, suspense file, retrieval and distribution of DISOSS/PS, so that the user can handle image documents as text and data documents, for example:

— store an image document in a DISOSS file with text and data documents on related subjects, entering indexing/filing information while the image is displayed on the screen of the workstation;

— make an image document, or part of it, available to application programs for use in an environment which does not support Scanmaster 1 directly.

The Wang PIC (Professional Image Computer) is based on the Wang Professional Computer (PC). The PIC supports existing PC system software, applications and data communications options; PC users can acquire image capabilities with PIC upgrade options. When configured with the appropriate communications options, the PIC can function as a workstation on a VS, OIS or Alliance system, accessing disks and printers attached to these systems. With the Remote Communications option, and appropriate software, a PIC can transfer images to remote PICs using the Systems Networking Point-to-Point Transport, whose capabilities include TeleSend, TeleMail and Mail Box. The PC software directly com-

patible with the PIC includes System Software with Interpretive
BASIC, Business Graphics, Multiplan, Word Processing,
Notebook (a sort of free-format database system) and Database
(menu-driven software for the management and interrogation/
retrieval of structured records and fields).

The software of the Philips Megadoc system is part of the Philips
file server Office Automation Monitor, which is based on the
P4000 DINOS operating system. The P5020 word processor may
be attached to a Megadoc system; one European installation uses a
Megadoc system for incoming correspondence, linked to a WP
system for outgoing mail. The WP and proposed teletex interfaces
use the T73 document storage standard for final form presentation
and an IBM DCA (Document Content Architecture)-compatible
option. DOCSYST, a document tracking system, will maintain a
document usage log and permit categories of documents to be
recalled satisfying, for example, such criteria as 'no answer
received during the previous two weeks'. This feature will help to
integrate the Megadoc system into business applications and procedures in offices, complementing the ISR (Information Storage
and Retrieval), which allows documents to be indexed and
retrieved.

Security

Security is the protection against loss of availability, integrity and
confidentiality from accidental and/or deliberate causes. In the
context of an image filing system, there are a number of security
considerations that may be relevant for a particular application.

Whereas a retrieved hard-copy document can be misfiled when
the borrower returns it, this situation does not arise when electronic information is retrieved.

An image can be simultaneously retrieved by several workstations, whereas a hard-copy document can become temporarily
unavailable to others when borrowed by one person.

Access to documents of different classes of confidentiality can
be controlled by labour intensive and costly manual procedures,
but access to images can be controlled more effectively and
economically by an automated image filing system designed with
document, file and log-on controls.

Hard-copy document back-up is expensive in terms of storage space and document handling costs, whereas image filing back-up can be provided more effectively and economically as protection against loss of originals from theft, fire, etc.

EASE-OF-USE CRITERIA

An image filing system, like a text filing and retrieval system or any other electronic office system, must be easy to use and fit into the office environment.

Then it will gain acceptance from its users and planned benefits will be obtained. Users will be on the look out for further uses to which it can be put and co-operate in the development of systems and procedures which integrate it with other electronic office systems.

Ease-of-use criteria are considered in the following categories:

— scanning and filing;

— retrieval and editing;

— transmitting and printing;

— system management and integration;

— ergonomic considerations;

— environmental considerations;

— user support.

These criteria are selected to highlight the components of the application of an image filing system as a user perceives them, whereas the functional criteria selected in the previous section highlight the basic functional components as a system designer perceives them.

Scanning and Filing

If the technology allows a document to be scanned in a few seconds, the user does not want to spend several minutes deciding how to set the controls and actually setting the controls, and moving the moving parts, every time a document is scanned. In practice, it is not as bad as that.

Relevant criteria are:

— the speed of scanning and filing;

— controls for setting up the scanner;

— document indexing and filing aids;

— menu-driven procedures.

Some documents will be in as good condition as brand new bank notes. Others will not. Curled corners, folds, tears, paper clips, staples, large size documents, pages of books and bound reports, etc may or may not need preparation before scanning and special handling during scanning, depending on the scanner itself as well as the application. In some cases, controls may need to be altered and the document rescanned. The time taken to scan and file a collection of documents of the various types and qualities that will occur when the image system is operational should be determined in trials, during which any difficulties can be noted.

In practice, scanning should be straightforward and the availability of aids to help produce a good digitised image will be appreciated by the user. It should be noted though that scanning a set of documents is not something that can be fairly casually rushed through like photocopying. Each document and the scanning system should be matched to each other, rather as a radiographer prepares a patient for an X-ray scan.

Whereas, in photocopying, no image of the document is stored, the indexing and filing (for retrieval and re-use) of the scanned digitised image of a document in an image system must be done and should be facilitated by software aids, perhaps using a menu-driven procedure and keying descriptive information and keywords into a form of options. Much has been written about text filing and retrieval systems, and is equally applicable to the filing and retrieval of image systems. In particular, the following advice is taken from pages 54 and 55 of NCC publication *Text Filing and Retrieval Systems* (see Bibliography, item 3).

'Where the system is intended for professional or administrative use it is important that the filing of documents can be carried out simply and efficiently. The greater the need to specify specialist computer details in order to carry out this function, the less likely the system is to be accepted by the user. In this respect two features can bring about benefits:

— the system can make use of function keys to aid the filing tasks. Function keys should be easily identifiable and can be used to perform a single task or be general purpose, with the action being dependent upon the menu currently displayed;

— formatted screens can simplify the filing of semi-structured documents. In such cases the insertion of control characters, necessary to identify individual fields, can be automated and the user prompted for the next field when actually inputting the text.'

At present, some image systems, although good in other respects, would benefit from further development of their indexing and filing software. In other cases, where the image capability is added onto existing electronic office software for text management, the indexing and filing software is good and easy to use. Depending on the application, it may be considered desirable, or essential, to be able to retrieve relevant image and text documents from an integrated database of both types of documents with a single retrieval enquiry. Consequently, users may expect to be provided with integrated facilities for indexing and filing both image and text documents. At least one supplier has such convenient facilities, whilst another hopes to provide these before the end of 1985.

Retrieval and Editing

Image retrieval, like text retrieval, should be and can be made simple for the user. This is achieved through the user carefully planning an electronic filing system which is appropriate for the application(s) and through the availability and usage of user-friendly software for indexing, filing and retrieval.

In a sense, image retrieval is simpler than text retrieval because the user does his retrieval upon the text descriptors of the content of the image only. He cannot yet in many systems directly interrogate the content of the image itself. But this might be seen as a drawback too. For it means that users must carefully index and file each image, whereas in some text retrieval systems there are facilities for retrieving documents simply by specifying words or character strings known to appear within their contents and in some instances this is a most convenient way of proceeding. In the 1990s, retrieval and other capabilities may be made very easy to use – one of the objectives of the research into and the development of fifth generation computer systems.

In some image applications, the main purpose of the user will be to retrieve and look at an image; in others, it will be to re-use the image – to manipulate, enhance or edit it. It should be simple for the user to know what he is trying to do and how to achieve it, and it should also be easy for him to achieve. Workstations like the Xerox 8010 (formerly called the Star) and the Apple Macintosh have dramatically made it easier for users to interact with a workstation and the information displayed on its screen and held on its memory. Through software and mouse manipulation, the manipulation, editing and filing of graphics and text is very easy, as most readers have probably witnessed at exhibitions, although they do not have these facilities yet on their own office workstation. A number of image filing systems have facilities just like these and the ease of electronic painting, with mouse-controlled manipulations, upon the display of an image, is very impressive. The reader is referred to Figure 1.5 and the icons displayed in the left-hand margin, which can be used in electronic painting.

In many office applications of image filing systems, the two activities which the user will frequently want to do should be simple to perform. The prospective user is advised to sit down at different systems and try them out for himself to find out. These activities are:

— windowing an area of the display and redisplaying it at higher resolution, ie zooming;

— manipulating the layout of a text document displayed on the screen and fitting images (of rectangular shape or, in some systems, of user-defined shape) into the document, ie electronic cut and paste.

Transmitting and Printing

Image filing systems do not give rise to any special ease-of-use criteria in relation to the transmitting and printing of images, over and above those ease-of-use criteria encountered in office systems that handle text.

It may be noted that one supplier is planning to increase the limitation on the maximum distance of the image printer from the control unit during 1985 from 200 metres to 1,000m.

System Management and Integration

As the image market is relatively immature, significant developments can be expected in the functionality of image systems and in the user friendliness of their functions, particularly in those functions integrated with other office systems, equipments, services and networks.

Ergonomic Considerations

Much has been written about ergonomics of VDUs, and other office systems, in particular pages 60-65 of NCC publication *Text Filing and Retrieval Systems* (see Bibliography, item 3), relating to screen, keyboards and printers.

Two factors of special interest to image filing systems, not applicable to most text systems, are mentioned here. First, an image system may have a mouse (a hand-operated electronic device for cursor movement and control) to assist the user in editing the image in various ways and to interact with the system. Second, the system will have a scanner, which a text system does not.

In some applications, the image filing system may be used rather like a photocopier, shared by many users who go to it now and again, use it and return to their desk. In other instances, it may be located at a user's desk. It may be an additional function which he can perform at his workstation, along with text document creation, text filing and retrieval, spreadsheet work, electronic mail and so on – or it may require a separate high-resolution display screen, in addition to his existing workstation. The point is that efforts are being made to reduce the 'footprints' of desktop workstations, ie the amount of desktop space they occupy. A mouse requires some additional space. This is not a criticism of mouse technology, and is probably not very significant when set against the advantages of a mouse, but is something to watch in the context of ergonomics, especially if the scanner itself is also taking up desk space.

If the user is continually spending a significant proportion of time creating mixed-mode documents, then attention must be given to the integration of the scanner into his desktop working environment.

Environmental Considerations

Most, if not all, of the hardware components of an image filing system will normally be located within the user's office environment, rather than inside a purpose-built computer room, where environmental factors can be controlled to match design specifications. Accordingly, for various environmental factors, the parameters within which the system is designed to be used should be found out, as should the range of values of these factors that will be encountered within the proposed user environment(s). Besides such factors as temperature, humidity and power supply, consideration should be given too to noise, size, heat output, etc.

For example, for one system, the specification is:

— scanner: 425 watts;

— display unit: 100 watts;

— control unit: 460 watts and 1,250 BTU/hour heat output;

— thermal printer: 200 watts;

— laser printer: 200 watts;

— temperature: between 60°F and 90°F (16°C and 32°C);

— humidity: between 20% and 80% relative humidity, non-condensing.

In another system, it is:

— scanner: between 32°F and 122°F (0°C and 50°C) for operating, and between 14°F and 167°F (–10°C and 75°C) for storage; between 5% and 95% relative humidity, non-condensing; convection cooling;

    — display unit: between 32°F and 104°F (0°C and 40°C) for operating, and between 14°F and 167°F (–10°C and 75°C) for storage; between 10% and 95% relative humidity, non-condensing.

Noise is unlikely to be a problem, but size and heat output may place constraints upon the location of the hardware components. There may need to be sufficient clearance from walls for air flow, whilst access requirements of maintenance/repair engineers must also be allowed for. As with any office computing equipment, consideration must be given to the quality of the electricity supply and written assurances sought from suppliers about whether the public supply is suitable or whether a dedicated power supply is called for. A point sometimes overlooked in the automated office is the need to deal with cables trailing on the floor. Good 'wire management' not only looks good, it promotes cleanliness and tidiness, and above all prevents personal accidents and equipment malfunctions caused by somebody tripping.

It is especially important in the context of image filing systems, where documents are scanned, that the lighting is suitable. Books and theses have been written about locating VDUs and word processors so that characters displayed on the screen can be easily and clearly read. This factor is of no less importance where images are displayed, especially when half-tone displays and image enhancement software are involved, or where one is trying to interpret an image of scribbled handwriting, a detailed drawing, etc.

But the scanner must be carefully located so that there is a controlled lighting environment where neither external natural lighting nor street lamps, nor internal lights, have an adverse effect upon the quality of the image. Remember that when you see a system in operation at a supplier's demonstration centre or at a customer site, the environment will look good, because it has evolved through planning, attention to detail and dealing with any adverse feedback from users.

User Support

Initial user training is necessary, but is not sufficient in itself to educate the user in all the operational aspects and fine details of an image filing system, or any other electronic office system. Therefore other aids should be available:

— user documentation;

— help facilities;

— technical enquiry service.

User documentation, the first level of user support, must be:

— available on, or before, the first day of installation;

— presented in a logical manner and carefully laid out so that it is easy to locate the information being sought (an index is helpful);

— concise and in clear English so that it is easy to understand (a glossary is helpful).

Thick documentation may be off-putting and virtually useless for many users, who are unfamiliar with much of the jargon and many of the concepts of computers and office technology. Documentation held on-line within the system can be easier to update and to access through a text filing and retrieval system which may be available at the display station of the image filing system (and indeed some systems have an integrated image database and text database), or can otherwise be made available at a separate microcomputer.

Help facilities allow the user to change from the particular procedure he is following in the system and ask for instructions and advice about what to do next. Not only is it important that the procedure used to call up the help facility is simple, but the instructions and advice that are presented to the user must be easy to understand, useful and relevant.

When the user cannot obtain the required information, there should be access to trained people who can deal with his problem. In the first instance the user organisation should arrange to have one person trained to have a good working knowledge of the system. For more complicated problems, the supplier of the system should provide a technical enquiry service giving advice on all aspects of the system.

SUPPLIER CRITERIA

Evaluation criteria presented in this section concern the suitability of the supplier of an image filing system and arrangements associated with the purchase. These are considered in the following categories:

— vendor characteristics;

— product package;

— product reputation.

Vendor Characteristics

It is always advisable to establish the reliability and dependability of the supplier you are dealing with. Not only do these characteristics give an indication as to the quality of the product being purchased, but they also indicate the future stability of the supplier. The latter point is significant since an image filing system could form an integral part of the development of office automation within the organisation. During the time that an organisation is using the system, support will be required from the supplier for a number of reasons: the user will want to ask questions about the system; maintenance will be required; the system may need enhancing to meet the changing needs of the organisation. Consequently, it is important that the supplier is stable enough to be able to meet the requirements in future years.

Three types of vendor characteristics are considered:

— profitability;

— organisation structure;

— research and development.

Profitability

A good way of assessing a supplier's recent success and short-term future is the profitability of the company. Taking the profit for the last few years, together with the trend, will give some measure of the supplier's growth and future stability.

Organisation Structure

The number of employees and company structure can give some indication of the supplier's credibility. To turn the salesman's promise into a working system the supplier must have enough staff and a good management structure to direct their activities.

Research and Development

The amount that the company has been spending on research and development each year, and the associated trend, can indicate how much importance the supplier places on long-term stability. The buyer should also enquire about the supplier's future plans for the product: when are new or modified products due?

Product Package

Whilst the facilities provided by an image filing system are of primary importance, there are other factors which will affect the success of the implementation. The following items need careful evaluation:

— delivery date;

— installation;

— user training;

— documentation;

— warranty period;

— maintenance.

Delivery Date

The vendor of the system should give a realistic delivery date and provide evidence that it can be met.

Installation

Ideally, the supplier should install the system as they are the experts in this field and will be aware of any potential problems. During installation a certain degree of disturbance may be unavoidable. To keep this to a minimum, the installation should be well planned and areas of responsibility clearly defined.

User Training

User training is vital and any training agreement with the vendor should clearly state:

— the content;

— who carries out the training;

— where the training is to be carried out (ie user or vendor site);

— how long is spent (ie vendor man days);

— whether there is any follow up.

The level of training required will vary from one individual to another. Some people will only use the system for very simple filing and retrieval tasks, while others may wish to utilise more complex facilities, such as image enhancement and electronic painting, or take advantage of associated features such as word processing, forms filling or electronic mail. In addition, system managers will need to be trained to provide all the support facilities for the general user.

Some systems may have training programmes on-line or self-teach packages. These can be particularly useful as they allow users to undertake the training at their own convenience. Such a facility can also be invaluable for training new employees as they join the organisation in the future.

Documentation

A comprehensive collection of documentation to meet the needs of all the users of the system should be included as part of the product package.

Warranty Period

When purchasing an image filing system, the buyer should insist on a warranty period to insure against those faults most likely to occur just after installation. A good way of defining the start of this period is the completion of a set of acceptance tests. These tests are mutually accepted by the supplier and buyer as an adequate method of showing whether the system is working to its specification.

Maintenance

Any system may, at some time, develop faults. These are often associated with hardware, although some subtle software faults may become apparent only after using the system for a long period. In addition, if the software is enhanced to improve it or to replace old bugs, new bugs may well appear until the software has become mature through exposure to users. Therefore, it is likely that some form of maintenance is required on the system. The usual means of providing maintenance is by contract with the supplier of the system. The cost of maintenance varies a great deal with the level of service provided. In general, the contract will cover replacement parts, labour, regular preventative maintenance checks and a guarantee to respond to a breakdown within a specified time. In certain circumstances, the response time for emergency service may be especially critical, since information held on the system will constantly be needed to aid decision-making processes.

Product Reputation

When assessing an image filing system it is worth establishing what reputation the product has with its other users. In order to establish this, the following should be investigated:

— quantity sold;

— other customers' opinions.

Quantity Sold

The first question to ask the supplier is "How many systems have been sold over what period of time?". If this number is small, the product and supplier should be viewed with caution. The product may be very new and is much more likely to present problems during installation than a well tried system with a large user base. An alternative reason could be that the system is not very good and other users have chosen a superior product.

Other Customers' Opinions

Other users can often provide a useful informed opinion of a product. In the first instance the supplier should be willing to supply user reference sites. However, when visiting other users, the buyer should bear in mind that users are not always willing to admit to having made a mistake.

Other users should be specifically asked about the reliability of the system. Once a system shows itself to be unreliable, it will not take long for the user population to lose confidence and stop using its facilities.

Since the UK image filing market is not mature, prospective users may wish to contact users in the USA or Western Europe.

3 The Evaluation Procedure

There are the following six steps in the evaluation procedure:

— establish your requirements;

— draw up a checklist;

— establish what products are available;

— select a shortlist of suppliers;

— assess the products;

— evaluate the products.

ESTABLISH YOUR REQUIREMENTS

It is important to be aware of basic requirements before undertaking the evaluation process. At the very least, the following information should be known:

— what application(s) the system is to be used for;

— what type and quantity of original documents are to be scanned and stored in the image database (eg typed and handwritten correspondence, pages of books and magazines, photographs, forms, etc);

— by whom the system is to be used and where they are located (nb the end users who access the image database and make use of filed image information, and the persons who possess and feed the original documents through the scanner may be different people and, indeed, work in different rooms or buildings);

— what computing, office technology and communications equipments, networks, services, facilities, etc the organisation already has and uses, or plans to have and use (nb these could be used in conjunction with – indeed may have to interface and interwork with – the image filing system).

DRAW UP A CHECKLIST

In order to evaluate the image filing systems that are available, you should specify which of the criteria described in Chapter 2 you wish to take into account. To help you in this exercise, the criteria are listed in the Evaluation Score Sheet (see Appendix A). First, make a copy of the several score sheets (the master score sheets at Appendix A should not be marked, as further copies will be required later), then work through the score sheets doing the following:

— strike out any criteria which are not relevant;

— identify those criteria which are absolutely essential for your application(s) by marking an 'E' on the left of the 'Comments' column;

— add on any additional criteria that you require in the spaces provided on the score sheets.

Those criteria which you have specified as essential, as opposed to desirable or irrelevant/unnecessary, will be used to establish a shortlist of suppliers. You can then proceed to investigate in detail the suitability of the products of these selected suppliers.

ESTABLISH WHAT PRODUCTS ARE AVAILABLE

In the course of providing this evaluation guide (in the early part of 1985), the following image filing systems were encountered:

Company	System
Advent Systems Ltd	Imager 20
Correlative Systems Int'l	VIPS 2000
Data Logic	Image System
Digithurst Ltd	MicroSight/MicroScale

IBM United Kingdom Ltd	Scanmaster 1/Image View Facility
International Computers Ltd	Document Image Handling System
**Kodak Ltd	**Kodak Image Management System (KIMS)
Philips Business Systems	Megadoc
Rank Cintel Ltd	Retriever
Sintrom Electronics Ltd	Datacopy
Techex Ltd	Computerised Archiving System (CAS)
**Toshiba Corporation	**Document Filing System DF-2100R
Wang (UK) Ltd	Professional Image Computer (PIC)

**Please refer to supplier for further information about whether this product is available in the UK.

The reader should be aware that the above list may be incomplete and, like any product list, can become out of date very quickly. Latest information may be sought directly from suppliers or from NCC's Information Service (061-228 6333, ask for the Enquiry Desk).

SELECT A SHORTLIST OF SUPPLIERS

The essential criteria that you have specified can now be used to build up a shortlist of suppliers. Take each supplier in turn and determine whether it can meet your essential criteria. It may be possible to undertake this exercise by using the information provided in the supplier's brochures, or it may be necessary to contact the supplier directly to ask specific questions. Brief comments describing how each supplier measures up to your essential criteria can be written onto the evaluation score sheets.

When you have completed this exercise for each supplier, you should select three or four suppliers who can best satisfy your essential requirements.

ASSESS THE PRODUCTS

Having established the shortlist of three or four suppliers, a detailed evaluation of their products may be undertaken. This can be done by using the brochures given to you by the supplier and by asking questions and requesting further information by telex, teletex, telephone, letter or electronic mail VANS. It is recommended that you see the products in operation, both at the suppliers' offices or exhibitions and at existing customer site(s).

In order to establish how the selected suppliers can best meet your requirements and at the same time determine detailed costings, an 'Invitation to Tender' should be sent to these selected suppliers. This should contain a specification of requirements including details of:

— document scanning throughput;

— number of scanning stations;

— number of high-resolution retrieval workstations (nb a retrieval workstation may double up as a scanning station);

— image printing throughput;

— number of image printing stations;

— storage requirements (nb different types of document give different compression factors, so estimate the number of documents of each type of content);

— transmission requirements;

— application(s) and interworking with word processing and other office systems, equipment, services, etc;

— retrieval response time;

— system expansion and interrogation plans.

It may also be useful to send each supplier a copy of the evaluation score sheets and ask them to provide information in the 'Comments' column.

Assistance in drawing up an 'Invitation to Tender' specification may be obtained from NCC's Advisory Services (061-228 6333, ask for Advisory Services).

When suppliers have submitted detailed costings this information can be used in the evaluation procedure and a Cost Summary Chart is included in Appendix B for the summarising of this information. It is suggested that costs are summarised under the following headings:

— hardware;

— hardware maintenance;

— software;

— software maintenance;

— other (eg training, installation, delivery).

The costs established should be exclusive of VAT and insurance.

In order to undertake a detailed assessment of the products on the shortlist, a set of the score sheets in Appendix A should be photocopied for each supplier. Each criterion can then be considered against each product, indicating if a product provides a particular function and, if so, to what extent. Any additional pertinent information can be added in the 'Comments' column. The measure of how well a product satisfies a given criterion can be indicated by allocating scores along the following lines:

Qualitative assessment	Score
No value: does not exist	0
Extremely poor	1
Poor	2
Weak	3
Below average	4
Average	5
Above average	6
Good	7
Very Good	8
Excellent	9
Perfect: cannot be improved	10

Having completed this exercise, you should have a completed score sheet for each supplier, with all the selected criteria scored. This will enable the next stage to be performed – that of comparing the suitability of the several shortlisted products.

EVALUATE THE PRODUCTS

Having completed the assessment exercise, it is necessary to determine which is the preferred supplier and to make a recommendation to whichever person or committee has delegated responsibility for authorising procurement of office systems within the organisation. Whatever the outcome of the product evaluation process, the procurement process in a particular organisation will require that the money can be spent. It may also be necessary for approval to be given separately, for other reasons such as:

— to ensure that the image system is compatible with the organisation's corporate information technology strategy, regarding integration with other systems and services, etc;

— to ensure that the internal and external auditors are satisfied that the system is secure and can be audited, wherever this is relevant, depending on the application(s);

— to ensure that the use of the system complies with an existing new technology agreement.

In many organisations, all these issues and the cost come together in the procurement authorisation process, but some may be overlooked in organisations where the investment in new technology and its application have advanced at a faster rate than have the administrative procedures.

One way of ensuring that the determination of the preferred supplier reflects the individual needs of an organisation is to adopt a simple weighted ranking method of evaluation. This technique involves assigning a weighting factor to each criterion to reflect its relative importance. A blank Evaluation Table for this exercise is contained in Appendix C.

Each criterion is assigned a hierarchical level, as illustrated below:

Level 1	1.1	Functional criteria
	1.2	Ease-of-use criteria
	1.3	Supplier criteria
Level 2	1.1.1	Document scanning
	1.1.2	Image storing and filing
	1.1.3	Image retrieval and display
	etc	(for functional criteria)
	1.2.1	Scanning and filing
	1.2.2	Retrieval and editing
	1.2.3	Transmitting and printing
	etc	(for ease-of-use criteria)
	1.3.1	Vendor characteristics
	1.3.2	Product package
	1.3.3	Product reputation

As many as are considered necessary of the level 2 criteria may be extended to a third level. As many as are considered necessary of the level 3 criteria may be extended to a fourth level, and so on. A basic hierarchical structuring, for image filing systems, is shown on the score sheets in Appendix A.

The next stage is to assign weights to each criterion. This exercise involves making a judgement on the relative importance of the individual criteria to your organisation for the application(s) for which the image filing system will be used.

In the first instance this involves assigning percentage weights to all the criteria at each level, so at level 1 this could be (nb the procedure is being illustrated and each organisation must decide for itself what values to give to the weights):

Functional criteria	60%
Ease-of-use criteria	25%
Supplier criteria	15%
Total	100%

At level 2, under functional criteria, percentages could be assigned as follows:

Document scanning	15%
Image storing and filing	10%
Image retrieval and display	15%
Image editing	15%
Image transmission	10%
Document printing	15%
System controller and integration	15%
Security	5%
Total	100%

This exercise should be completed for each level within each of the categories – Functional, Ease-of-use and Supplier criteria. Figure 3.1 shows an example of how percentage weights are assigned, and how actual weights are determined, in a hypothetical evaluation exercise.

The percentage weights that have been allocated at each level are used to calculate the ACTUAL WEIGHT factors to be allocated to specific criteria. Actual weights should be calculated for all the criteria which require answers about a supplier's product.

Take, for example, the criterion 'Is the scanner able to accept original documents of the various sizes that will arise?', which appears in Figure 3.1 and under Functional Criteria in Chapter 2. This is a level 3 criterion and an aspect of Document scanning (level 2), which in turn comes under the level 1 heading Functional criteria. The ACTUAL WEIGHT for this criterion is calculated as follows:

— the weight for Document scanning (level 2) is calculated as a percentage (15%) of the weight for Functional criteria (level 1), ie 15% of 60 = 9.00;

— the actual weight for the 'size of document' criterion (level 3) is calculated as a percentage (10%) of the figure just calculated, ie 10% of 9 = 0.90;

— the value 0.90 is now entered in the ACTUAL WEIGHT

column against the 'size of document' criterion and is used to determine the weighted score for each supplier for this criterion.

Using the blank evaluation table in Appendix C the evaluation exercise can take place carrying out the following steps:

— using copies of the blank evaluation table, list all the criteria being considered in the evaluation, again marking 'E' against those criteria considered essential. Figure 3.1 shows how the completed table may look;

— assign percentage weights to each level of the criteria;

— calculate the actual weights for those criteria asking questions about the product (ie those criteria at the end of a branch of the hierarchy);

— transfer the scores for each criterion from the suppliers' score sheets (Appendix A) to the evaluation table in the appropriate supplier column;

— calculate the weighted scores, by multiplying the actual score by the corresponding actual weight, for each criterion and for each product;

— add up the weighted scores in each column to obtain the total weighted score for each product;

— look up the total cost (in £) for each supplier from the cost summary sheet (Appendix B) and insert these figures in the appropriate columns at the bottom of the evaluation table.

The cost/performance ratio can now be derived. This ratio is recorded in the evaluation table in Figure 3.1 as a points per £1,000 figure. This is arrived at by applying the following formula to each supplier's figures:

$$\text{points/£1,000} = \frac{(\text{grand total weighted score}) \times (1,000)}{\text{total cost (£)}}$$

The resulting ratios indicate who is the supplier offering the best 'value for money' product.

In certain circumstances, where specific financial regulations or

LEVEL	CRITERION	Weight At Level 1	2	3	4	Actual Weight	SUPPLIER A SCORE	SUPPLIER A WTD SCORE	SUPPLIER B SCORE	SUPPLIER B WTD SCORE	SUPPLIER C SCORE	SUPPLIER C WTD SCORE	SUPPLIER D SCORE	SUPPLIER D WTD SCORE
1	Functional	60												
2	Document scanning		15											
3	Type of document			10		0.90	6	5.40	9	8.10	8	7.20		
3	Size of document			10		0.90	7	6.30	9	8.10	8	7.20		
3	Nature of content			15		1.35	8	10.80	9	12.15	8	10.80		
3	Size of scanned area			10		0.90	8	7.20	8	7.20	8	7.20		
3	Resolution			15		1.35	8	10.80	9	12.15	8	10.80		
3	Number of bits per pixel			15		1.35	8	10.80	9	12.15	9	12.15		
3	Controls for setting up scanner			10		0.90	8	7.20	8	7.20	7	6.30		
3	Size of scanner			5		0.45	6	2.70	7	3.15	5	2.25		
3	Speed of scan			10		0.90	5	4.50	6	5.40	6	5.40		
2	Image storing and filing		10			6.00	9	54.00	6	36.00	7	42.00		
2	Image retrieval and display		15			9.00	9	81.00	6	54.00	7	63.00		
2	Image editing		15			9.00	8	72.00	8	72.00	8	72.00		
2	Image transmission		10			6.00	8	48.00	7	42.00	6	36.00		
2	Document printing		15			9.00	8	72.00	7	63.00	8	72.00		
2	System controller & integration		15			9.00	8	72.00	6	54.00	5	45.00		
2	Security		5			3.00	7	21.00	7	21.00	6	18.00		
	TOTAL FUNCTIONAL WEIGHTED SCORE							485.70		417.60		417.30		
1	Ease of use	25												
2	Scanning and filing		15			3.75	6	22.50	7	26.25	6	22.50		
2	Retrieval and editing		20			5.00	8	40.00	7	35.00	6	30.00		

Figure 3.1   Example of a Completed Evaluation Table

LEVEL	CRITERION	Weight At Level 1 2 3 4	Actual Weight	SUPPLIER A		SUPPLIER B		SUPPLIER C		SUPPLIER D	
				SCORE	WTD SCORE	SCORE	WTD SCORE	SCORE	WTD SCORE	SCORE	WTD SCORE
2	Transmitting and printing	10	2.50	7	17.50	5	12.50	5	12.50		
2	System management & integration	15	3.75	8	30.00	5	18.75	4	15.00		
2	Ergonomic considerations	10	2.50	7	17.50	6	15.00	6	15.00		
2	Environmental conditions	10	2.50	8	20.00	7	17.50	7	17.50		
2	User support	20	5.00	9	45.00	8	40.00	7	35.00		
	TOTAL EASE OF USE WEIGHTED SCORE				192.50		165.00		147.50		
1	Supplier	15									
2	Vendor characteristics	30	4.50	9	40.50	7	31.50	6	27.00		
2	Product package	40	6.00	7	42.00	9	54.00	8	48.00		
2	Product reputation	30	4.50	7	31.50	9	40.50	7	31.50		
	TOTAL SUPPLIER WEIGHTED SCORE				114.00		126.00		106.50		
	GRAND TOTAL WEIGHTED SCORE				792.20		708.60		671.30		
	TOTAL COST £				25,000		21,000		19,500		
	POINTS/£1,000				31.69		33.74		34.43		

Figure 3.1 Example of a Completed Evaluation Table (continued)

restrictions occur, it is accepted that procurement decisions may need to be made on purely financial grounds – selecting a cheaper system rather than the one which offers the best value for the money invested. There are two points to bear in mind, however:

— irrespective of the price, the system should always satisfy those criteria deemed to be essential;

— the cost of any proposed system enhancements, subsequently, may favour a system which does not have the cheapest outlay initially; in other words, a relatively cheap system may not have flexibility for expansion designed into it, whereas a more expensive system does!

When considering the points/£1,000 ratio, two or more suppliers may have very close or identical ratio figures. In such a case, a final decision could be made by examining how well each product scored against the essential criteria.

Appendix A

Evaluation Score Sheets

SUPPLIERS NAME:

PRODUCT:

PAGE	LEVEL	CRITERION	SCORE	COMMENTS
31	1	*Functional Criteria*		
32	2	Document scanning		
	3	Type of document		
	3	Size of document		
	3	Nature of content		
	3	Size of scanned area		
	3	Resolution		
	3	Number of bits per pixel		
	3	Controls for setting up scanner		
	3	Size of scanner		
	3	Speed of scan		
38	2	Image storage and filing		
	3	Storage medium and its capacity		
	3	Location of storage medium		
	3	Compression algorithm and speed of compression		
	3	Indexing the image		

PAGE	LEVEL	CRITERION	SCORE	COMMENTS
41	2	Image retrieval and display		
	3	Retrieval of image		
	3	Display modes		
	3	Display options		
44	2	Image editing		
	3	Image manipulation		
	3	Electronic cut and paste		
	3	Image enhancement		
46	2	Image transmission		
	3	Local options		
	3	Remote options		
47	2	Document printing		
	3	Resolution		
	3	Speed		
	3	Modes		
	3	Paper handling and other characteristics		
49	2	System controller and integration		
52	2	Security		

PAGE	LEVEL	CRITERION	SCORE	COMMENTS
53	1	*Ease-of-use Criteria*		
53	2	Scanning and filing		
	3	Speed of scanning and filing		
	3	Controls for setting up the scanner		
	3	Document indexing and filing aids		
	3	Menu-drive procedures		
55	2	Retrieval and editing		
	3	Retrieval		
	3	Windowing		
	3	Manipulation		
	3	Cut and paste		
	3	Other editing facilities		
56	2	Transmitting and printing		
57	2	System management and integration		
57	2	Ergonomic considerations		
	3	Mouse		
	3	Scanner		
58	2	Environmental considerations		
	3	Temperature		
	3	Humidity		
	3	Power supply		
	3	Noise		
	3	Size		

PAGE	LEVEL	CRITERION	SCORE	COMMENTS
	3	Heat output		
	3	Location		
	3	Wire management		
	3	Lighting		
59	2	User support		
	3	Initial user training		
	3	User documentation		
	3	Help facilities		
	3	Technical enquiry service		
61	1	*Supplier Criteria*		
61	2	Vendor characteristics		
	3	Profitability		
	3	Organisation structure		
	3	Research and development		
62	2	Product package		
	3	Delivery date		
	3	Installation		
	3	User training		
	3	Documentation		
	3	Warranty period		
	3	Maintenance		

PAGE	LEVEL	CRITERION	SCORE	COMMENTS
64	2 3 3	Product reputation Quantity sold Other customers' opinions		

PAGE	LEVEL	CRITERION	SCORE	COMMENTS

Appendix B

Cost Summary Sheet

	SUPPLIER A	SUPPLIER B	SUPPLIER C	SUPPLIER D
HARDWARE COSTS				
HARDWARE MAINTENANCE COSTS				
SOFTWARE COSTS				
SOFTWARE MAINTENANCE COSTS				
TRAINING COSTS				
INSTALLATION COSTS				
DELIVERY COSTS				
TOTAL COST				

Appendix C

Evaluation Table

LEVEL	CRITERION	Weight At Level 1 2 3 4	Actual Weight	SUPPLIER A		SUPPLIER B		SUPPLIER C		SUPPLIER D	
				SCORE	WTD SCORE	SCORE	WTD SCORE	SCORE	WTD SCORE	SCORE	WTD SCORE

Appendix D

Useful Addresses

Advent Systems Limited
12 The Business Centre
Molly Millar's Lane
WOKINGHAM
Berks RG11 2QZ
(Tel: 0734 784211; Telex: 858893)

Correlative Systems International NV
Parc Scientifique d'Evere
Rue de Strasbourg
1130 Brussels
BELGIUM
(Tel: 010-32-2-242-5700; Telex: 63598 COVIPS B)

Data Logic Limited
29 Marylebone Road
LONDON NW1
(Tel: 01-486 7288)

Digithurst Limited
Leaden Hill
Orwell
ROYSTON
Herts SG8 5QH
(Tel: 0223 208926)

IBM United Kingdom Limited
Contact local sales office, see telephone directory

International Computers Limited
118/128 London Street
READING
Berks
(Tel: 0734 586244)

Kodak Limited
(PO Box 66)
Kodak House
Station Road
HEMEL HEMPSTEAD
Herts HP1 1JU
(Tel: 0442 61122; Telex 825101)

The National Computing Centre Limited
Oxford Road
MANCHESTER M1 7ED
(Tel: 061-228 6333; Telex 668962 NCCMAN G)

— NCC Advisory Services (for advice)

— NCC Enquiry Desk (for information about suppliers
  and general information)

— NCC IT Circles Administrator (for information about
  any of the 5 Information Technology Circles covering
  Communications, Data Processing, Microsystems,
  Office Technology and Security)

Philips Business Systems
Elektra House
2 Bergholt Road
COLCHESTER
Essex CO4 5BE
(Tel: 0206 575115; Telex: 98673)

Rank Cintel Limited
Watton Road
WARE
Herts SG12 0AE
(Tel: 0920 3939; Telex: 81415)

Sintrom Electronics Limited
Arkwright Road
READING
Berks RG2 0LS
(Tel: 0734 875464; Telex: 847395)

Techex Limited
5B Roundways
Elliot Road
West Howe Industrial Estate
West Howe
BOURNEMOUTH
Dorset BH11 8JJ
(Tel: 0202 571181; Telex: 41437)

Toshiba Corporation
Tokyo
JAPAN
Enquiries to:
Lion Office Equipment
International House
Windmill Road
SUNBURY-ON-THAMES
Middlesex TW16 7HR
(Tel: 09327 85666; Telex 889219)

Wang (UK) Limited
Wang House
661 London Road
ISLEWORTH
Middlesex TW7 4EH
(Tel: 01-560 4151; Telex: 8954121)
(or contact local sales office, see telephone directory)

Appendix E

Bibliography

1 *Introducing the Electronic Office*, S G Price, NCC Publications, 1979

2 *Electronic Indexing and Hard-Copy Management*, J A T Pritchard, NCC Publications, 1983

3 *Text Filing and Retrieval Systems – A Practical Evaluation Guide*, S J Newton, NCC Publications, 1983

4 *Introducing Electronic Filing*, P A Wilson, NCC Publications, 1986

5 *Office Printers – A Practical Evaluation Guide*, M A Condon, NCC Publications, 1982

6 *Business Graphics*, J M Bird, NCC State of the Art Report prepared for members of the Office Technology Circle, 1985

7 *Introducing Electronic Archiving*, J A T Pritchard, NCC Publications, 1985

8 *Graphics and Image in Office Systems*, S J Newton, NCC Publications, 1985

Appendix F

Glossary

Bit mapped   A raster scan display is refreshed continuously from a digital store that contains a digitised map of the scanned document (or edited image). Such a display is said to be 'bit mapped' (and the process 'bit mapping'). Most displays in terminals and word processors designed for text (alphanumeric characters) are character mapped.

Half tone   A term used in photography. A half-tone picture/photograph is one that is prepared, using appropriate technology, to give shades of grey between white and black. In image systems, an original document may have a continuous spectrum of tones from white through all shades of grey to black. Obviously such a spectrum cannot be represented precisely when the original document is scanned and digitised. A compromise has to be made in order to keep within reasonable limits the storage requirements, the speed of operation (scanning/compression and decompression/presentation) and cost of an image system. This compromise is determined by the user application(s) – as are the scanning resolution and other functional design characteristics – and so there are different image systems on the market designed for different market sectors.

                      For a given scanning resolution, the amount of detailed information which is stored in digitised form about an original document is determined by the number of bits used to represent each point. Some systems give the user a choice between one and eight bits per point: in the former case a particular point is recorded as black or white, and in the latter case any of 256 shades of grey (including black and white) can be recorded. When more than one storage bit per scanned point is used (this is sometimes referred to as the 'multi-valued pixel mode of operation'), the capability exists for the system to provide half-tone presentation. The design of the system's software and hardware determine how half-tone presentation is carried out, and the degree and manner of user/system interaction involved. The derived half-tone display or printout does not have the 256 shades of grey (or whatever other number has been recorded), and can be varied by the user adjusting the white/grey and grey/black thresholds.

Half-tone presentation is a binary represen-tation giving a pseudo-grey level effect by varying pixel size and density.

(The subject of colour, so important in the context of business graphics systems, is a very specialised application of image systems and is beyond the scope of this guide, which is con-cerned with systems designed for, or which can be used for, office applications.)

Interlaced refresh   See Raster scan display

Multi-memory plane   Used in the context of bit mapping. When each scanned point is recorded as either black or white, and one bit per point is used to store this information, a 'one memory plane' is said to be used to control the display on the screen. To handle shades of grey (and colours in busi-ness graphics systems), multivalued pixels arise and several bits (eg 3 or 8) are needed to record the state of each scanned point. Con-sequently further memory planes are needed, hence the term multi-memory plane storage of the scanned document.

Pan

A verb meaning to move the display on the screen. It is not normally possible (see Resolution) to present the image of an entire document upon the screen without loss of detail. When a portion of a document is displayed, with full detail, the user can pan across the stored image and cause other portions of the document to be displayed. The term scroll is also used. Panning may be controlled by a mouse, cursor keys or by some other mechanism.

Pel

See Pixel.

Picture element

See Pixel.

Pixel

The terms pixel, picture element, pel are frequently used by authors/lecturers in the context of image processing, facsimile, graphics, OCR and micrographics. It is beyond the scope of this book to discuss alternative definitions. The terms mean whatever their user intends them to mean, eg the smallest area or point of a document that is individually scanned, or the smallest area of a bit-mapped display that can be individually addressed. Standards bodies will define these terms in certain contexts, which may eventually clarify the situation.

Raster scan display

Raster scan technology has been employed in television applications for 50 years. An electron beam moves across the screen horizontally from left to right, starting at the top left hand corner and progressing vertically downwards. There are two modes of raster scan display: interlaced and non-interlaced. In interlaced mode, all odd numbered lines of the scan and all even numbered lines of the scan are refreshed with alternate sweeps of the electron beam, ie lines 1, 3, 5, 7 ... etc are scanned, followed by lines 2, 4, 6, 8 ... etc. In non-interlaced mode, every line of the display is refreshed on every sweep/pass of the electron beam. Very high refresh rates are used with raster scan, reducing the likelihood of flicker. An image is formed on the screen when the intensity of the beam varies during the scan. Raster scan displays have become the industry standard for business graphics.

Resolution
There are three uses of this term in image systems: scan resolution, display resolution, and print resolution. Frequently the term resolution is used on its own, so the context determines the type of resolution. In business graphics systems, the term resolution is used without qualification on its own and without confusion or ambiguity. It is, perhaps, rather unfortunate that the term should so often be used without qualification in the context of image systems.

In business graphics systems, the concept of scan resolution does not of course arise. Since the printed output is derived from stored text, data, and special graphic elements, and from relationships between these items of stored information, the concept of print resolution and how faithfully the printed output represents/matches an original document does not arise either. Resolution refers to screen resolution and is expressed as m × n, or as m points × n points, or as m points per horizontal line × n horizontal lines, or as (mn) points. Displays of up to 375 × 375 (ie 140,625 points) are regarded as low resolution, displays in the range 500 × 500 to 1,000 × 1,000 are regarded as medium resolution and are found in good business graphics systems, while high-resolution displays of up to

4,000 × 4,000 can be provided for CAD/
CAM, engineering and other specialist applications. For example, a resolution of 512 ×
512 means that this number of discrete points
(262,144) can be displayed and addressed.
This requires 32K bytes of storage for black
and white only.

In image systems, scan resolution refers to
the number of points per inch (eg 200 ppi ×
200 ppi) in each of two perpendicular directions in which the original document is scanned, ie it refers to the fineness or coarseness of
the scan. It is also expressed as m points per
inch by n lines per inch. This occurs because a
scanner scans a line at a time. Scan resolution
may also be expressed as a single quantity, eg
200 or 200 ppi. This is interpreted as being the
scan resolution in each of two perpendicular
directions. However, it should be noted that
the resolutions in the two directions may be
different. Scan resolution would not, though,
be expressed as (in this instance) 40,000
points per square inch. Scan resolution is also
sometimes expressed as points per millimetre,
eg 8 by 7.7, or 8 by 8 in conformance with the
CCITT Group IV facsimile standard. Similar
comments apply to print resolution, where
again the user is interested in fineness/
coarseness of presentation. Print resolution
need not necessarily coincide with the scan
resolution from which printouts are derived.
If the printer operates in a finer/coarser mode
than the scanner, software can transform the
stored bit map appropriately to match it to the
printer.

Display resolution, in image systems, is
expressed in two different ways which is con-
fusing. It may be expressed in the way,
described above, used for business graphic

systems, eg as 800 by 600 (ie 480,000 points per screen). This describes how much information can be displayed. It may also be described, as scan resolution and print resolution are, in terms of the density of presentation, eg 100 ppi by 100 ppi.

For technological reasons, it is realistic to provide higher scan and print resolutions than display resolution. Typically, these may be respectively 200 ppi, 300 ppi and 100 ppi. Consequently, the original document cannot normally be displayed in its entirety on the screen of the retrieval workstation with the level of detail captured during scanning, because of the reduction in density and because the screen and the original document have different rectangular dimensions. An image of the entire document may be displayed with a loss of detail and reduced dimensions, or a portion of the document may be displayed with no loss of detail but with increased dimensions (often superimposed as a window upon a coarser display). An original document may be scanned with about four million points recorded, whereas the screen may be able to display only about an eighth of this number of points at any moment. Some systems are using a single resolution (200 by 200) for scanning, display and printing, but market forces (ie user requirements and product costs) may give rise to other configurations.

Strictly speaking, display resolutions should quote the screen size to be meaningful.

Scroll   See Pan

**Vector refresh
mode**   In business graphics systems, vector refresh mode is an alternative method to raster scan

for producing a graphical display on the screen. In vector refresh mode, the electron beam can be moved simultaneously in both the X (horizontal) and Y (vertical) directions to produce straight lines (or arcs), as a pencil is used to draw a picture.

Window   Used as a verb and noun. A window (usually rectangular) is identified upon the display, in preparation for some manipulation, such as:

— in a text document, a window is created for the insertion of a filed image; the text layout is automatically rearranged (this is referred to as 'electronic cutting and pasting');

— in an image document, a window is created and the contents are displayed at full detail, superimposed upon the current background display.

Zoom   Used as a verb. Zooming (or windowing) defines an area of the displayed image which is then re-presented in full detail (ie the area is viewed at actual bit map resolution).

Appendix G

Workshop Reports

During 1983, NCC held 'Graphics and Image in Business' workshops in London and Glasgow. Some of the conclusions reached by the syndicates will be of interest to readers of this book and are reported below.

Issue 1:   How will image processing change the working practices within the office?

Report: — For the purpose of the report, image processing was defined as "taking a snapshot of a document – which remains constant".

                — It will enable documents containing images, pictures, etc that have been generated externally to be captured and held electronically.

                — Communications will be improved, in that document transfer will be speeded up and users will have the ability to transfer documents containing foreign languages, symbols, diagrams, etc.

                — The fact that a document is held electronically in a central location will result in many users having immediate access to the latest version.

                However, it was agreed that before image processing could be fully integrated into the electronic office, a number of problems must be ironed out:

                — there must be international standards to cover the capture, transmission and retrieval of images;

                — an image processing system should incorporate a high-resolution screen for viewing the captured documents and a quality printer for obtaining hard copy as required;

                — the question should be resolved of whether a signature contained within a document that is stored electronically is legally binding and if that document is an acceptable substitute for the original hard copy;

> — the stored image should be able to be edited and manipulated in such a way that it can meet the cutting and pasting requirements of the publishing area;
>
> — external documents captured by an image processing system should be able to be held together with internal documents generated electronically to form a "case file";
>
> — there should be long-term plans to integrate image with text, voice and data.

Issue 2:   What will be the major advantages of storing, retrieving and/or transmitting images within the office?

Report of Syndicate A:

1 The ability to electronically store documents containing text and image will reduce the need for retaining hard copy.

2 Holding text/image documents on a central database will make them immediately available (if desired) to a larger number of users.

3 Electronic mail can be utilised to speed the transfer of digitised text/image documents to remote sites.

4 The electronic storage of digitised images can ease the problem of merging documents containing diagrams, sketches, etc.

5 The storage of text/image documents in a digitised form will give benefits in the area of security – both in terms of access and back-up.

Report of Syndicate B:

1 The combination of machine-readable information and image together in the same electronic document is a very real benefit. It should improve presentation and also give the user the option of viewing either at the terminal or in hard-copy form.

2 The electronic storage of text/image documents will enable multiple access – two or more people can have simultaneous

access to the same document.

3 The ability to have multiple access means that there will be considerable time savings in the circulation of papers, when compared to the traditional methods such as distribution lists, etc.

4 Storing documents electronically will reduce the need for hard copy and should result in office space, previously used for physical filing, being made available.

5 Statistics such as the frequency of retrieval and usage are much easier to obtain when documents are held in electronic form.

Applications

1. Solicitors
2 Insurance claims
3 tech documents
4 correspondance
5. land registration diagrams
6. Mag disk ⇒ optical.
7 Librarys
⇒ pg 22 food.
1 GB = 1 000 000 000 bytes.

8. Rental Agreements.
9.

1. Scanning
2. Storage + filing = Indexing
3 Retrival + display
4. Editing.
5 transmission
6. Printing

7 System Mangement + Integration
8. Ergonomics
9. environmental
10 user support.

Index